How to Kill Bed Bugs

Bed Bug Treatments and Tips for Killing and Getting Rid of Bed Bugs

By Hamon Ilyde

How to Kill Bed Bugs
Bed Bug Treatments and Tips for Killing and Getting Rid of Bed Bugs
By Hamon Ilyde

Printed in the United States of America

Copyright c 2010 Hamon Ilyde

LEGAL NOTICE

Table of Contents

Introduction

Bed bugs are not something that most of us really want to talk about. In fact, there are few things that are this worrisome to talk about. Yet, it is essential that homeowners are aware of this threat.

You may think of a bed bug as just something from a riddle from when you were a child. Often, our parents would say to us, "Don't let the bed bugs bite." Yet, the fact is that this can and does happen today.

Bed bugs are small creatures that are hard to see and hard to notice. Yet, their presence is something that will cause you not to sleep well at night. Often, the thought of bed bugs can send chills down our spines.

So, as a homeowner what should you do if you feel that you may have this infestation in your home? What should you look for when you visit a hotel or stay at someone else's home? A bit of education on this subject matter really can help you to know what to expect, what to look for and what to do about it.

Bed bugs are growing in population around the world. Today, there are more and more of these little creatures coming back into homes and this is somewhat of a worry. Just like a pesky ant or fly, you need to work at getting rid of these creatures so that you and your family remain safe.

What You'll Find Here

In our ebook, you will find a great deal of knowledge that you need to have. All in all, you will learn about these creatures including things like where they came from, what they are all about and what they are doing in your bedding.

It is also important that you know what to do if you think you may have bed bugs in your home. We will provide you with a step by step look at how to rid your home of these awful little guys.

Bed bugs can be frightening and worrisome, but that does not mean that you can not get rid of them and keep them gone.

Taking the first step in that search is to educate yourself about this species and then to learn what you can do to protect yourself and your loved ones from it.

<u>Chapter 1</u> - What Are Bed Bugs?

Bed bugs. You have heard the word but do you really know anything about this creature? Most people do not. Many do not even realize that they actually exist. But, they do and they may be lurking in your home, in your bedding or even in your carpeting.

It is essential that you take the time necessary to learn more about these pests. By doing this it will help you to succeed in treating them, and having a home that is free from an infestation of the worst kind.

While some of the information you will read here is a bit graphic, you will learn from it just why it is so important for you to rid yourself and your home of these nasty little creatures.

What Are They?

The bed bug is a creature that likes to feed off of human beings. It is his food of choice. If they can not find a person to feed from, they will select other warm blooded animals to use as a host for their necessary feeding. This can include your pets like cats and dogs. It also includes birds, rodents and bats as well. Their food is that of the host's blood.

With that being true, you usually find these creatures in locations were there is food for them. They are found in homes and in other locations where there is a good amount of hosts. This would include in places like hotels, motels, shelters, apartment buildings, dorm rooms at colleges, and even prisons. Anyplace that there is a good source of human hosts, in a large turnover amount, is a good place to find bed bugs.

How Do They Get There?

If you have a home, you may be wondering how these bugs have gotten into it. One of the major ways in which bed bugs do infest is through means of transportation that we all use. Often, they can invest methods of transportation such as busing lines, trains, and various types of passenger and commercial ships and even in the airlines.

They can literally infest these locations because there is a large amount of hosts located in them. Because there is usually a large amount of people in any of these places at any given time, they make for the ideal location to be if you are a bed bug.

So, how do they get from these transportation units into your home? Bed bugs are able to be transported from any of these transportation methods to your home in several ways.

- They can be transported through the clothing of their host person.
- They can be transported through the luggage that comes from an infested location.
- They can come through furniture that is carried on board one of these units.
- They can come from bedding as well.

There are plenty of ways in which the bed bug can make his journey from one place to the next. Often times, people have no idea that the location they are staying in, such as a hotel room, is infested. They then come home; luggage packed, bringing the bed bugs with them.

Within a matter of time, the bed bugs have grown in their numbers and can easily move from one place to the next. Soon, there are large numbers of them. They will leave you wondering just where they came from in the first place.

Hygiene and Bed Bugs

One thing to note is that if you have an infestation of bed bugs in your home, this does not mean that you are a bad housekeeper.

Bed bugs are very versatile and strong creatures that are more than capable of hiding and lurking in places where you may not ever look.

In fact, they would prefer to live in a home that is clean anyway. That means that if your home is clean, you may still have the presence of bed bugs in it.

Also, it does not mean that you have poor hygiene yourself if you have bed bugs in your home. Again, bed bugs do not want to live and host from an unclean source, although they will if they have to.

You do not have to be a dirty person or live in a dirty home to have an infestation of bed bugs in your home.

Later in our ebook, we will talk about what the bed bug does and does not react to. It is essential that you know now, though, that most bed bugs are not going to respond to the same types of treatments that you give to ants or other pests that infest your home.

Remember, they use their host for feeding. They need blood to be sustained. Therefore, they are not going to necessarily respond to food products that you put out.

Before you begin to worry if you have bed bugs in your home, take the necessary time to insure that these little creatures are not there by knowing what they look like, what they do and what possible clues you have.

Bed bugs may be something that is a bit frightening, but you can learn what to do to treat them. And, although you may not want to hear it, it is likely that the bed bug will be found in your home at one time or another. Therefore, if you do not have an infestation right now, you should take note of what can happen if you do.

<u>Chapter 2</u> - What Do Bed Bugs Look Like?

Now that you know a bit more about this bad creature, it is time to learn more about what they actually look like.

You will find several pictures located here to help you, but it is important to read the necessary information about them as well. Bed bugs are commonly mistaken for other pests in the home. Or, you may think that you have bed bugs in your home when you actually do not but have some other pest lurking there.

If you plan to treat bed bugs, you need to have proper identification of these insects. Here are some key points to help you to identify them.

- In their adult stage, they are brown to a reddish tint of brown.

- They are an oval shape but they are also flattened out.

- In size, the adult bed bug is about 3/16 to 1/5 of an inch long.

- When they have just had a meal, the bed bug is swollen looking. They will be longer now and have a dark red color to them.

- On the front of their heads, they have what looks like a beak that allows them to pierce and suck from their mouths.

- Adult bed bugs do have wings. These wings do not allow them to fly, though. They are small in size and are very short looking.

- The eggs of bed bugs are white or colorless. They will darken in their color as they mature. They will eventually have a brownish tint to them when they are mature.

- The nymphs look a lot like that of the adult bed bugs in their appearance, just at a smaller size.

They Look Like Others

Bed bugs are commonly mistaken for other bugs that are in the same family, (Family Cimicidae) or in a closely related one. They are often confused with Cimex adjunctus, which are bat bugs or with Cimexopsis spp which are chimney swift bugs. The swallow bug, Oeciacus spp is also confused with bed bugs quite often.

Often the only way to know if a bug really is a bed bug is to have a professional look at it. A microscope is used to determine if the pest has the tell tale characteristics of a bed bug, as we have mentioned above. It is often necessary for a skilled entomologist to do this type of investigation as most can not tell the difference from these creatures listed to the bed bug.

Pictures

Here are some pictures of bed bugs that can often be used to help distinguish the look of a bed bug:

Here is a picture of a bed bug. In this picture, the bed bug has not eaten and therefore has it's tell tale flattened look about it.

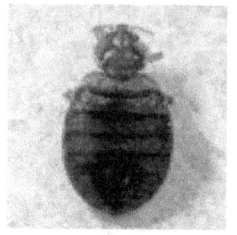

In this picture, you can see the size difference between a bed bug that has eaten and one that has not. As they eat, they become elongated. They are also swollen and are therefore thicker.

In the following picture, take the time to notice what the bed bug looks like in various stages of its life. The bed bug is commonly found in the same area at all these stages. If you have adult bed bugs, you will also have eggs. This is what makes it hard to treat the bed bug, because as you may kill off the adults, the eggs may still be present.

The Life Cycle of the Bed Bug

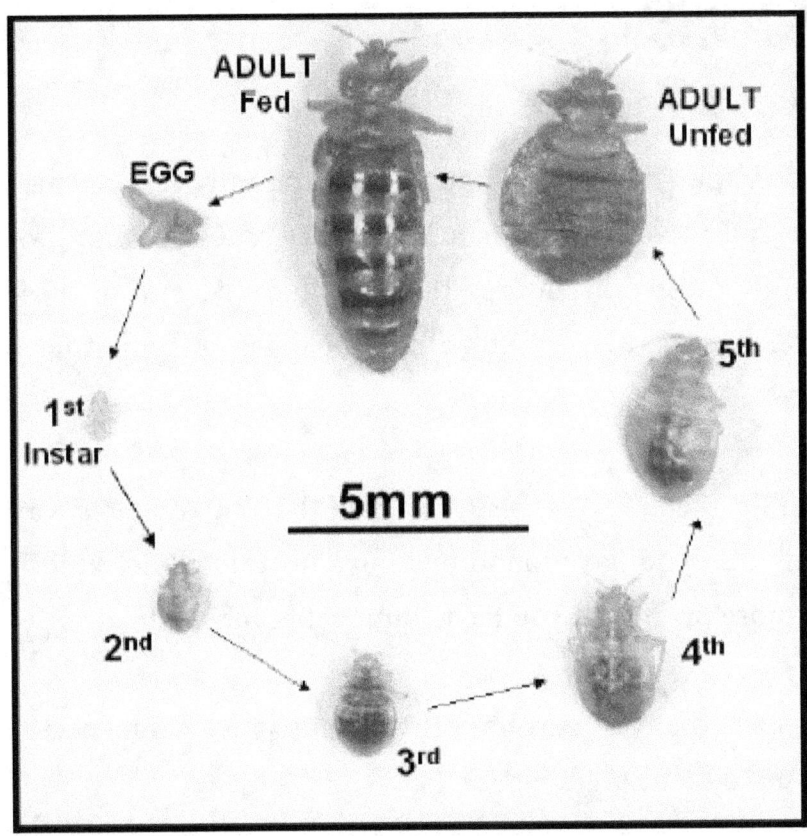

In this final picture, you can see just how small a bed bug really is. This will show you just how hard it can be for you to notice that they are there.

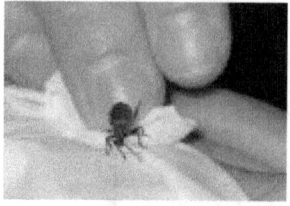

Remember, though, that it is quite hard to really tell if a pest that you find in your home is actually that of a bed bug. For example, if you live in an area that may have several of these closely related species; you may find out that you do not have bed bugs after all.

For example, if you live in the Midwest of the United States, you may think that you have found bed bugs in your home. But, it is much more likely that you have found a bat bug instead.

If you suspect that you may have them, contact a local inspector to help you to identify them.

<u>Chapter 3</u> - Where Are Bed Bugs Found?

Now that you know what the bed bug looks like, you need to determine where it is that bed bugs can live.

Like all animals, bed bugs prefer certain things about the locations that they live in. While they may be able to live just about anywhere, these creatures do have preferences.

Bed Bug Species

There are two types of main bed bug species.

- Cimex Lectularius: This is the common bed bug that you are most likely to find.

- Cimex Hemipterus: This is known as the tropical bed bug because of the areas that it prefers to live in.

Where They Are

You can commonly find bed bugs living throughout the world. Yet, as we mentioned they do enjoy specific climate regions.

- Cimex Lectularius, the common bed bug is commonly found in areas that are cooler in temperatures. They are found in areas of North America, in areas of Europe and in Asia as well. In these areas, the bed bug tends to be in the cooler regions as apposed to the warmer or tropical regions.

- Cimex Hemipterus, this is the tropical bed bug. It is commonly located in warmer temperatures. It is a sub form of the other in which the bed bugs have adapted to the warmer temperatures. This tropical form of bed bug is found throughout areas such as in the tropical climates. They are found in Africa, Asia and throughout North and South America, in the tropical regions in these areas. This particular bed bug is only found in the Florida region of the United States

Are You Safe Then?

Just because you live in one of these areas does not mean that you have to have bed bugs in your homes. As you will see later on in our ebook, there are things that you can do to prevent them from getting into and infesting your home and property.

Likewise, this does not mean that if you do not live in one of these areas mentioned that you can not have bed bugs in your home. It is very common that bed bugs will spread from one location that is infested to another that is not by several means.

They can easily move from one location to the next just as you would. They move from one place to the next through transportation means that you use.

For example, if you visit Florida this summer for a wonderful vacation, you could potentially bring the bed bugs home with you in your luggage, on your person or in your clothing and bedding products. They can survive a good amount of time without feeding so you may not even notice them on your clothing when you get home.

What is important to know is that the bed bug truly can move from one location to the next without being noticed. As we talked about before, the bed bug is a very small insect and one that is likely to go unnoticed. It can do this easily because it is so small and flat. It can hide in the small cracks and crevices throughout your clothing or luggage.

A tropical bed bug from Florida probably can breed in other areas of the country, depending on the climate and the warmth of that location. So, even if you do not live in one of these areas in which bed bugs are commonly found, that is not to say that you can not bring them back to your home from them.

Other Specie Infestations

There are several other types of bed bugs that should be mentioned. These are from the same family as the bed bugs that we have talked about, but they are a bit different in their reactions and where they will live.

- **Leptocimex Boueti**: In this form of bed bugs, the bed bug is found in the tropical areas of West Africa as well as being found in South America. This bed bug will also infest bats as well as its human preferences.

- **Cimex Pilosellus and Cimex Pipistrella** are two forms of bed bugs that generally infest bats throughout the bat population in virtually all areas in which the bat is found.

- **Haematosiphon Inodora**: This is the type of bed bug that is found in North America. Yet, this bed bug prefers to invest poultry instead of infected humans.

- **Oeciacus**: This is not completely that of a bed bug, but shares many of the same characteristics and is commonly thought of as a bed bug. In this species, the pest will infest more birds than it will the human population. It can be commonly found throughout locations in which there is a large bird population.

While these other forms are located around the world, they limit their connection with humans to that which is just necessary. Some species will use a human host or any other warm blooded creature that they can get into contact with and the location of them does not matter, then.

More to Know About Infestation

There are many places that the bed bug is lurking. You may be thinking that perhaps they are only found in your bedding, but this is very much not true. In fact, the bed bug is a very defying creature and is more than capable of being where you do not know it to be.

Size wise the bed bug is very small and so flat that it can fit into a location that is just wide enough for a credit card. A space that thin is large enough for them to crawl through.

During The Day

The hardest time to find bed bugs is during the day time. You will find them hiding in small, hidden locations. Here are some places to look for them:

- Mattress seams and interiors, usually they are completely out of sight here.

- In furniture, especially furniture that is near the bed or in the bedroom.

- The bed frame can house them during the day time hours too.

- Carpeting that is plush may provide enough space for them to hide.

- Baseboards or in other products found in the bedroom.

- Picture frames, inside of books, inside telephones, inside curtains...these are all good places that they would hide.

- Sofas and other areas in which you spend a decent amount of time can harbor them.

- They can hide under wall paper that has been loosened, in the cracks in the plaster on your walls, under wall coverings of all sorts, and even in your ceiling molding in your home.

- Check the electrical boxes, your door frames, your windows and your window frames.

- If you have wall to wall carpeting, do not forget to get under the carpeting to look for them. Do not stop there; they are under the tack board under your carpet too.

As a general rule of thumb, the bed bug is likely to venture upward, outward and virtually anywhere that it can easily move and hide behind. They like to be out of sight and out of light, so look for them to hide.

The heaviest infestations can allow them to be seen more often. Of course the last thing you want is to have more bed bugs to deal with! In this case, they can be seen in larger groups and are often more visible because of this.

Generally speaking a bed bug will stay within 100 feet or so of its host. Usually it feeds when the host is lying still such as at night. They only need to feed once every five to ten days, so this allows them to travel from place to place between feedings. They can be found virtually any place in the home, office or other dwelling.

How Many Are There?

Learning that you have bed bugs is something that is very troubling. Most people will think that they just have too many mosquitoes around, but the fact is that there may be bed bugs lurking about.

If you know that you have bed bugs in your home, you now need to determine how many you have so that you can properly handle them.

Unfortunately, there is not one sure fire way to know that. The biggest factor in determining how large your infestation is will be to determine, or try to determine, when the time of the initial infestation happened.

If you remember from earlier, we talked about how the bed bugs have gotten into your home. They probably have come in through you or someone else bringing them in while they were traveling. But, you should also remember that no matter where you may have gone, as long as you go to a location that is infested, you can bring them home.

This means that you can just head to a neighbor's home that is infested and they may come home with you. This is especially true if the bed bug infestation is very large. It is much more common that they will be lurking in clothing that is packed in luggage, though.

So, How Do You Know?

It is very common for bed bugs to reproduce quickly, which you will learn more about in a few chapters. But, what is important to note is the possibility of more than one infestation in the home.

For example, you may have your initial infestation in your bedroom, in your bed. Perhaps you brought them home from a trip you took in the luggage. You have no idea that they are there. But, they are. This would be your main and first infestation area.

Now, what if your child is sleeping in your bed? They can possibly carry the bed bugs back to their bed as well. Now, you have a secondary location of an infestation of bed bugs in your home.

It is essential that you seek out help for finding all of these hidden infestations and get treatments for them right away. This is the best method of preventing a very large infestation that is harder to control.

One Note

While bed bugs can be located in virtually any area of your home, they are most commonly found close to their host. The host, of course, is you or your family. While they can be in any area of the home, it is most common to find them in the bedroom area and likely on or in the bed itself.

But, if you have an infestation in your bed, you will want to check out the rest of the home as well to prevent further problems from occurring. Look in areas such as your living room, your carpeted areas, your other beds and bedroom furniture. Make sure to look in areas that are dark, cracks that are evident and in small crevices that exist. You are not likely to find them outdoors, but they can be hiding in your pet's bedding as well.

A thorough search should include all of these areas of your home.

Chapter 4 - A Brief History of the Bed Bug

You may not think of an animal like the bed bug of having a history, but this one does. In fact, it has been found that this is one of the strongest species of animals present on the Earth because of what it has accomplished.

While most humans would be okay with the bed bug being extinct this is anything but what is likely to happen over the next years.

A Look Back

Taking a look back, it can be found that the bed bug has been found in ancient writings. It is commonly believed that the bed bug has been around for thousands of years, probably as long as humans have inhabited Earth as well.

The bed bug was first called a pest in the 17th century. They came to the Americas just as the colonists did. They traveled with them on board the shipping vessels that were used by immigrants. With each new wave of colonists came a new wave of bed bugs.

The 1940's and 1950's

In the 1940's and the 1950's, the United States was using a product called DDT. This is known as the first type of pesticide that was used. It was first developed in the early period of World War II. Its purpose was to stop the spread of infectious diseases such as malaria and typhus by killing off the mosquitoes and other pests that helped to spread the disease.

DDT was used throughout the military but it was also used by everyday people as well. This allowed it to be used heavily and virtually everywhere in the United States readily. It was even used as an agricultural insecticide.

So, what does this have to do with the bed bug? DDT actually almost caused the extinction of the bed bug in North America. During the middle of the 20th century, it was hard to find a bed bug.

Are They Gone, Then?

Unfortunately for most people, the bed bug did not die out during this time period but over the last several decades has staged a come back that is large enough to cause them to re-infest many areas quickly. North America has seen a tremendous growth of bed bugs in the last ten or so years.

But, there is something different this time around....

DDT is no longer used today as a pesticide. It was banned from use, as were other products that are similar to it, as it was found to be dangerous not only for the bed bug and other pests but also to the human population itself.

This poses as bit of a struggle for those that are trying to treat bed bug infestations today. Many of the products that are used today are just no effective at treating these pests. In fact, the bed bug can be just as hard to treat and get rid of as that of the cockroach. In short, there is no 100% guaranteed method to treat an infestation of bed bugs.

More Reasons

There are additional reasons that the bed bug has made somewhat of a comeback over the last few years. This reason has to do with the type of products being used to treat pests. Because of all the health risks of DDT, many products have a lower toxicity level than those that were once used.

They are also more of a gel based product. Pests are less likely to respond to the gels but have been shown to respond to the spray products that are again, not in use.

On top of all of these things, the bed bug has developed, big muscles, so to speak. In fact, the bed bug has learned to adapt to these insecticides and therefore they are almost immune to them. This is due to the fact that many of these pest control products were used so heavily to treat other insects that they have learned to just adapt to them.

In fact, today, there are no gel based insecticides that do actually have any effect on the bed bug. These are the most common types of insecticides that are used in the industry today.

Feeding Traps

With some other types of insects that have developed immunity against sprays and gel based products, the most effective treatment tool is that of the use of food baits. The insect is lured into the trap because it smells of a product that seems to be that of food. The pest enters the bait trap and carries the food back to his colony. The food is usually poisoned.

This is the most effective way to treat animals such as ants and other hard to kill insects. But, can you guess what happens with the bed bug here?

The bed bug does not and can not respond to these types of traps. There is one simple reason for this. The bed bug does not feed off of those types of products. The bed bugs main food source is that of blood, in most cases, human blood. Therefore these types of traps are ineffective at treating bed bugs.

In 2005 and 2006, there has been a lot of attention paid to the bed bug. The goal is to educate those that have the potential for exposure to the bed bug to know about it. Because bed bugs have not been around heavily since before 1940, most people do not know what they are, where they are or what they are capable of doing.

The Professional Pest Management Association is an advocacy group in the United States that is providing the campaign for this information to be presented to the general public in a way to educate them about this possible infestation.

Why Know History

It is important to know the history of the bed bug for a number of reasons. Like any other type of history education, we can learn from the mistakes and the trails and errors of the past.

With bed bugs, it is important to notice where they lived, how they spread as well as what they did to the public.

By looking at the use of DDT and other chemicals, we can better see what will work on the bed bug, but we can also see the effects of harmful products on the human population as well.

The Future

As you look at the past of the bed bug, you should take note of the likely future as well. It has been shown that the bed bug is likely to continue in their growth and infestations. With more and more people traveling from place to place on airplanes, it is even more likely that the bed bug will go with them.

For this reason, it is important for you to know when you may have a possible infestation and to help you to get rid of them; you need to know how to do so.

Luckily, you will learn that here.

<u>Chapter 5</u> - The Life Cycle of the Bed Bug

Like all animals, the life cycle of the bed bug is a very normal thing. They are born, grow and live. But, as those that are trying to get rid of a bed beg will tell you, it is very important for you to have an understanding of what this lifecycle is so that you can better handle your infestation problem.

It is unlikely that you will actually learn when your first infestation happened. This is unless you know exactly where they came from. But, if you assume the time frame from which the first infestation came, you can get a better idea of just how potentially large your infestation actually is.

Size Matters?

When it comes to learning the size of the infestation that you have, you will want to try to get an estimate. This will help you to get a good handle on what exactly you can do about it. While it may upset you to learn how large your infestation really is, knowing can be the first step to getting rid of them.

Why The Lifecycle Matters

We are going to touch on the basics of the life cycle of the bed bug. We do this so that you have a better understanding of where they are within your home. By taking the time to learn where the bed bugs are in their cycle, you can provide the appropriate treatment to get rid of them.

The Life Cycle Of The Bed Bug: Broken Down

The female bed bug is the main focus from the start. She will lay up to twelve eggs per day! While it may only be one egg, it can be as many as a dozen.

She will place these eggs in a specific location and generally it will have the same characteristics. It is usually a rough type of surface. Or, she may place them into a small crack or groove that she finds.

She can place them on any rough surface because of the coating that is on them. It is sticky and will pretty much stick to anything.

It can take from six days to seventeen days for the eggs to hatch.

The baby bed bugs are called nymphs. Their first order of business is to feed. In fact, they are able to feed from a host right after being hatched.

They need to find their first meal quickly as this blood meal is needed for their development. They will molt after their first blood meal.

They will go through a total of five cycles of molting before they will reach their adult size.

In most cases, from start to finish, the egg will go from being a small egg to being a full grown adult in as little as 21 days.

Temperature Matters

One thing to take note of when looking at the life cycle of a bed bug is the temperature in the area. In order for the egg to hatch, the nymph to molt and for the bed bug to grow, the right temperature must be met. This temperature needs to be between 65 degrees and 86 degrees.

If the temperature is not met, the maturity of the bed bug is usually delayed. If the temperature is at 86 degrees, the bed bug will mature in about 21 days. If the temperature is that of close to 65 degrees, it can take as long as 120 days for the maturity of a bed bug to actually happen.

Food

From the time that the egg hatches, the main goal of the nymph will be to find food. It needs a blood meal in order to grow and to receive its first molting session. From the time it hatches until it reaches adulthood, it needs to molt at least 5 times. Yet, it can not do this without the right amount of food.

Again, this period of nymph will likely be elongated if there is not enough food to provide the necessary growth and molting periods for the bed bug. The time that it takes for the nymph to reach maturity has a direct relation to the amount of food that it has.

How Long They Live

Unfortunately for the human population, the bed bug is likely to make it through its nymph period and into adulthood.

One reason for this is as simple as the fact that they can live several months without eating any food at all. While they like to feed every five to ten days, they can survive several months without any food whatsoever.

Once they reach their adulthood, the female bed bug will begin to reproduce.

Once they reach their adult life, the bed bug is likely to live between a year to a year and a half. This is dependant upon on how much food they receive.

Finally, the bed bug can produce eggs at least three times per year. More often is also common.

When you factor these things together, you can see just how large an infestation can be when it has just occurred only a few months before.

<u>Chapter 6</u> - The Habits of the Bed Bug

As someone that is looking to get rid of bed bugs, it is important to understand just what these creatures like to do. The habits of bed bugs may amaze you even though you really do not like them.

The bed bug is not a creature any of us really want to get to know, but understanding more about them will help you to get rid of them.

Eating

The bed bug likes to eat at night. He is a nocturnal blood feeder. He likes to find his host when it is sleeping and therefore lying still. This poses less of a risk for his well being, of course.

They are very quick moving animals and can easily get out of the way if needed.

They will use their very sharp, pointed beak to break the skin of the host. By piercing it, they open it up enough to insert a fluid within it. This salivary liquid is what allows them to withdraw blood from their host. It is what is called an anticoagulant which will stop the host's blood from clotting and closing up the pierced area too quickly.

An adult bed bug will take up to fifteen minutes to feed from that one pierce. Most adults will be filled within ten minutes but can eat for up to fifteen.

The nymphs, or babies, will start to feed as soon as they are hatched and can find food or a host. When they do, the can only feed for as little as three to four minutes. As they grow, they will feed longer until they reach their adult size.

The bed bug does not need to eat very often. It can go several months without consuming any food. But, it will usually begin to seek out the host again after five days for another feeding. The bed bug will look for his host again when he is hungry as it will take him this long to digest the blood meal he has just consumed.

What They Like

Do you care what bed bugs actually like? Most of us would say no, but we still should take a look at this topic for a better understanding of these little creatures.

First off, they enjoy the dark and prefer not to come out unless it is dark. They are creatures that like to hide and stay out of sight.

To hide, they will find small crevices and cracks to hide in. This may be places like fabric or wood, but any place will do as long as they are hidden.

Usually, the bed bug will not travel too far from its host, as it wants to stay close to its source of food. But, they can and do venture away if they so choose to do so. Although they are small, they can travel throughout the entire house if they wanted to do so.

Yet, it is most common to find them near their host's bed or in the small vicinity of where they know they can find their host.

The most common location for them is in the folds of a mattress.

Searching For Bed Bugs

When you decide to go on a search for bed bugs, you will want to try to look for them at night. This is the most common time that they will come out to look for their host. Remember that they do not feed everyday so you are not likely to see all of them at once.

Still, you can do a bit of searching for them during the day; it is just harder to find them this way.

Odd Facts About The Bed Bug To Know

- Bed bugs are attracted to the carbon dioxide that humans put out when they breathe. They are also attracted to the warmth that a person provides.

- If you are bitten by a bed bug, you are likely not to feel anything for at least several hours. It will take this long for your skin to react to the products that the bed bug will inject into you.

- Some scientists believe that bed bugs can go as long as 18 months without eating any blood at all. Although this is not common, it is thought to be possible.

- In her lifetime, the female bed bug is likely to lay about 500 eggs. This is only if she bears just five eggs per day.

- When reproducing, the battle is on. The male bed bugs will actually impale the competing males. In doing so, he will inseminate the other male. Then, when the other male tries to inseminate the female, he will actually inseminate the female with the other male's secretions instead.

- Although bed bugs do not like to be outdoors, they can be. They can even climb up under wood and other materials to invade another home. This is common in condos and in apartment complexes.

- You may see a bed bug trapped inside the weaves on a woven shirt or pair of pants. They will appear as a dark speck there and are often missed by viewers.

Chapter 7 - What Can A Bed Bug Do To Me?

While all of this is great, what you want to know is what the bed bug will end up doing to you.

The good news is that bed bugs are not that dangerous to most people. While no one wants to have them around, they are not likely to provide you with any real problems, although in some people they can cause a higher level of reaction than others will have.

Bed bugs are often thought of in the minds of children as biting. In the sing song, "Don't let the bed bugs bite," they may have described just what the human can expect from the little bed bug.

The good news is that it is painless to the human. The bed bugs will likely feed from their host at night, while the host is sleeping and still.

Therefore, it is likely that the host, or human, will never feel it or will they see the bed bug actually bite them.

The bed bug will inject a liquid into the bite that it creates. This fluid is used to keep the blood from clotting and sealing up the wound. In most cases, the only way that you will know that you have been bit is by the reaction that some people have to this fluid.

If you do have a reaction to the fluid, it is likely to be something that is bothersome but not overly problematic. It will be an itchy, irritated and inflamed area of the skin.

But, each person is different here. Many people will not have any reaction to the bug bite at all, which makes it harder to know they are even there.

Harsher Reactions

Some people will react in a much more severe way. There is no way to tell which way you will react, until you do. If they react severely, it is likely that they will have a swollen and hard area on their area of their skin at the bite mark. It will be small and usually looks like a white welt on the skin.

If you have this serious of a reaction to the bug bite, you will likely have a great deal of itching to go along with it.

If you have serious reactions to the bed bug bite they can last from just a few hours to days. It is wise to seek out the help of a professional when you have these serious reactions to any type of bug bite.

One of the most common signs of bed bug bites is having three or more bite marks or welts in a row. This is thought to happen because the bed bug will become detached to the area, possibly by the host moving, and will then need to open a new piercing to draw blood from.

One thing to note about the bed bug is that they are different from flea bites. They will not have a red dot in the middle of the bite mark, but will be a solid color throughout the mark.

It is important to notice any marks that you may have that are small, bite like marks that you really can not explain.

This can be a sign that there are bed bugs lurking in your bed at night.

More Side Effects?

Bed bug bites do not commonly cause serious reactions in humans. But, they are annoying enough to cause stresses in our lives as well.

Some people that have experienced bed bugs in their home have problems with sleeping in their beds. They may feel anxiety and worry about the bed bugs, even knowing that they are not going to cause a great deal of damage to them.

This fear can lead to insomnia as well.

At an extreme level, bed bugs can cause things like distress and alarm. At some times, it can also cause delusional parasitosis as well. Here, a person becomes delusional, thinking that they see and feel the bed bugs on their skin.

The amount of blood that you will lose because of the bed bug is a very small amount and will likely not affect the host at all. They probably will not notice any feelings caused by this problem.

Also, bed bugs are not known to cause any type of disease. They are not known to carry and transmit diseases either. This is unlike that of a mosquito that can transmit a disease to its host.

They have been known to carry what is called pathogens for the plague and for hepatitis B, but this is just from the immunities that they have built up from these diseases over the course of time.

The bed bug bite that does cause a serve reaction or one that leaves a welt on the skin can become infected if the individual scratches at the area and by doing this, supplies the necessary bacteria to cause the infection in the skin of the host.

Chapter 8 - How to Know You Have Them

Before you move on to how to get rid of the bed bug from your home, we want to stop for a moment and determine just what you need to do to know that you actually have them.

Now, you will recall that it can be hard to know if the pest infesting your home is actually that of a bed bug. While only a professional that looks at them closely will be able to tell you right off if they are bed bugs, it is pretty much sure that if you have these signs that you are infested with bed bugs in one from or another.

Signs You Have Bed Bugs

Here are some things to look at to consider if you do in fact have bed bugs in your home.

- Obviously, if you see a bed bug then you would likely know that there is more than one there. You would likely need to look for the bed bug to actually spot it, though.

- You may see blood stains on your bedding. This is usually caused by the bed bug being crushed while it is on the host. For example, you may role over onto the bed bug while it is feeding.

- Brown or reddish brown spots on the bedding or clothing can be signs as well. These are the fecal matter of the bed bug. Excrement can be found on the sheets, mattress, on the walls, or on other areas that they travel.

- If you find things such as eggshells or shredded skin on your flooring, bedding or on your mattress, this can be a sign of bed bugs. These are generally found on locations that are close to their hiding places.

- If you have a very large infestation in your home, you may have an unexplained odor in the home. This will be a sweet yet musty smell to it. Usually, it is offensive and is noticeable when you come into your home.

- It is normal to just have one bite mark on the body, but a tell tale sign that you have bed bugs is the look of three bite marks in a row. In fact, these are often caused by the movement of the bug in a line as it looks for the same blood vessel it had before it was disturbed.

- A great way to notice them is to use a method that will allow you to detect their presence. To do this, simply plan to turn on a light without moving much from your bed just a few hours or so before dawn. At this time of the day, the bed bugs are most actively seeking their host. Use a flashlight to catch them in the act or on the bedding.

Set A Trap!

If you really want to see for yourself that there are bed bugs in your home or in your bed, you can set a trap to do so. Here are some things that you can do to see their presence in your home.

- Place sticky tape, sticky side up on the area around your bed. As the creature crawls over it, they will stick to it. This will not necessary stop all of them, though.

- Place heating pads in an area that is dark. This is especially effective when you have used the heating pad in the past. Use tape to catch them or you can just watch for their presence.

- You can use either of these methods and combine them by placing a balloon that has been filled by mouth in the area. Remember, bed bugs are attracted to the carbon monoxide that people exhale. This will help to lure them in.

A trap like these can help you to lure in the bed bugs so you can actually see if you have an infestation. Of course, there is no guarantee that the bed bug will in fact respond to these types of traps.

Store Bought Products

There are also products on the market that you can purchase that can help you. These are generally marketed as products to help you to get rid of bed bugs. Whether they work for that reason is another story. But, they can definitely help you to know if you have bed bugs in your home.

These products are quite like that of flea traps where they will lure the pest into the trap and are supposed to keep them there. Unfortunately, most of these traps are really only good for determining if you have bed bugs in your home. Most are not nearly effective when it comes to ridding the home of the pests.

Other Signs

All of the things that we have talked about thus far in the ebook are good signs that you have bed bugs. If you notice them, think you notice marks on your skin or on your children's skin, then you may have an infestation of bed bugs in your home.

Those that travel or have visited homes in which there are an infestation should take care as to determine if they too have brought the bed bugs into the home.

Beyond a doubt, the best way to know if there are bed bugs in your home is to use a professional service. They will come out and look for the tell tale signs and being experts will be able to tell you for sure what they have found.

<u>Chapter 9</u> - What to Do If You Have Bed Bugs

Now, we have talked quite a bit about the bed bug but what you want and need to know is what to do if you have discovered that you have bed bugs in your home.

We spent that time talking about it so that you would fully understand what these creatures do, how they live and how they affect you. Knowledge is power and in this case, it can help you to get rid of bed bugs once and for all.

The Steps to Getting Them Out

We will provide you with a step by step approach to methods that can help you to get rid of bed bugs once and for all. If one method does not work, move on to the next one. The goal here is to determine what the best method for your infestation is.

<u>Step One: Identification</u>

The first and most important step in treating bed bugs is to determine that the bed bug is actually a bed bug. As we mentioned, bed bugs are commonly mistaken for other creatures.

The problem here is if you treat them as you would other creatures you will have no success with treatment.

Likewise, if you treat another animal with the treatments recommended for bed bugs, you are likely to have few results as well. A flea will not respond to the treatment of a bed bug. And a bed bug will not respond to the treatment of a flea.

Step Two: Determine Method Of Treatment

The next thing that you need to do is to consider if you plan to treat the bed bugs yourself or if you plan to call in a professional to handle the situation.

If you have a serious infestation of bed bugs, it is likely that you do need to use a professional to handle the infestation if you will have any chance of getting rid of them in the long term.

Let's take a look at the options when it comes to each of these treatments.

Do it Yourself Treatments

You do have some options here. One method that can be effective is to use a mixture of pyrethrums and fresh water diatomaceous earth. These products will work to provide you with a natural solution to getting rid of bed bugs.

It will cause damage to the nervous system of the bed bugs in the form of a mechanical action. Then, the bed bug will die.

It is essential for you to not use these types of products that are salt water diatomaceous earth as this can be quite dangerous. In fact, any animal that inhales it, pets and humans alike, can have lung damage.

In addition to this, this product has been found to cause cancer in mammals of all types.

If you plan to use this type of product, there are some manufacturers that do provide an insecticide mixture like this for you to use. If you purchase it, you will want to insure it is used with the utmost respect for the directions provided with it.

Another option for the self treatment of bed bugs is to use fruit and vegetable insecticides. These will do the same as above but will also provide the necessary safety to other animals including humans.

These products are made of canola oil and pyrethins.

A Mistaken Self Treatment Approach

One treatment of bed bugs that you may have heard about is using extreme levels of heat and cold to kill them. As you have learned, bed bugs do prefer a specific heat range. They like the temperature to be above 65 degrees but below 86 degrees as well.

Yet, there is scientific proof that disproves the theory that bed bugs will in fact be killed by these extremes of temperature. The main problem is that there is not enough time to keep them in this state.

The product needs to remain very cold or very hot for a long period of time. This is just too hard to do, in most cases.

On top of this, there is no real way to do this to treat mattresses or other large bedding items. Therefore it is recommended that this method of removing bed bugs not be used.

Professional Help

Those that have bed bugs may not even want to do anything but call on a professional to come out and help them to get rid of the pests.

When you call on an exterminator to handle your bed bug problem, it is very important that you take the time necessary to seek out one that truly knows what he is doing.

You need to insure that he knows how to handle and has been effective at the treatment of bed bugs, not just any type of insect. They require special treatment options.

You may find that there is a problem with doing this. If you live in North America, the near extinction of bed bugs in this area has caused there to be very few people available that do know how to treat this condition.

As bed bugs begin to re-infest this area, though, it is a sure thing that there will likely be additional individuals to call on.

Another problem that may arise when calling a professional to handle the bed bug problems you are having is the fact that many of the techniques used to treat bed bugs when they were everywhere before the 1940's are no longer allowed to be used.

For example, Cyan gas was used to fumigate areas that had infestations. It was quite a powerful insecticide and it did work to destroy them. But, the risks that it presented to pets and the human population (not to mention the environment) have caused this substance to be no longer in use.

Fumigation may be the only way to go, but it is also costly and may be banned in some areas. You will need to insure that the exterminator that handles bed bugs in your area is well aware of these problems.

Finding a skilled, experienced bed bug exterminator is an essential part to getting rid of the bed bugs.

Now, we will say that you do decide to call on a professional to handle your extermination. This is recommended as it is the most effective.

But, don't worry, there is more that you can do to handle this situation.

Step Three: Treatment Needs

You should know from the start that bed bugs are not usually going to be gone in the blink of an eye. In fact, they can take several treatments in order to be effective. Often times, it is necessary that there are three or more treatments necessary.

The more invasive the infestation is, the more necessary it will be for them to do more treatments. The size of your home or location to be exterminated also should be taken into consideration here.

Things like having a skilled exterminator and having an extensive treatment procedure done can help you to get the most out of the treatments.

Additionally, it is also beneficial if you have an exterminator that will remove nesting areas from the home and help you to determine where the infestation came from.

When all of these factors are taken into consideration, the bed bug infestation can be treated faster.

You should take the time to determine the extent of what the exterminator will provide for you, in writing before you move forward with their service.

Step Four: Pre Treating Your Home

Once you have an exterminator in place, he should help you to follow the needs provided here. But, you will need to handle the preparations for the extermination yourself.

If this is not done right, the whole chemical treatment that you get will be compromised.

So, let's take a look at what you can do to protect your home and prepare it for extermination of the bed bugs.

- **Packing up the home**. The first thing that you need to do is to pack up the home. This will include moving your furniture away from the baseboards and walls. You will then need to open up anything that is potentially a hiding place for the bed bugs to expose it to be sprayed.

  You should empty out desks and drawers. Bookshelves should be emptied too. Anything that you pack up should be cleaned and treated for bed bugs. The last thing you need is to re-infest your home with bed bugs once you clean them out.

- **Washing**. The next thing that you need to do is to launder anything that can be washed. This is no easy task of course. You will want to make sure that the laundering will include things like your clothing, your rugs and even any stuffed animals that you have.

  To make this effective, wash and immediately place all items into a plastic bag and seal them. They should be removed from the house until after the treatment has been done.

If you will be using a dry cleaner, tell them of the situation and provide all laundry in sealed bags to them.

The most important part of the laundering of these products is that you allow them to dry in the drier for at least 20 to 30 minutes on high to medium heat. This should kill all bed bugs at all stages of their life cycle.

You do not even have to wash the clothing, although most will want to just to feel better about wearing them again.

- **Vacuuming Them Up**. This too is an important step in getting rid of bed bugs. Here, you need a high powered vacuum and you need to do a very detailed job of cleaning.

It will help to lower the number of bed bugs from the home, making the chemical treatment that much better for you.

You should use attachments to get into the crevices, the walls, the baseboards and virtually any place that the bed bugs are lurking. You can also use this on your mattress and bed frame as well.

Get behind pictures, under furniture and throughout the furniture to get up as much as you can. Remember all the places we said they liked to hide? Use that page of the ebook to help you to vacuum them up.

Finally, make sure that you take the filter and vacuum outdoors and empty it right into the trash. You do not want to give them any chance to escape on you.

- **Steaming**. Steaming can be done, but it will not necessarily work magic on your furniture, your mattresses or your bedding. Sometimes it can work well, especially if you do not want these items to be treated chemically, but in the long term, it is more effective to allow chemical treatment if the product will handle it.

- **The Bed**. One of the biggest decisions that you will need to make is what to do with the bed. If your bed has been infested, especially if it is a heavy infestation, you may want to consider getting rid of it. But, it can be treated, in most cases.

In most cases, local treatments using insecticide products will be effective. If the box spring is compromised, though, this can be hard to treat as there are many places for them to hide.

The decision is yours. If you feel that you will feel better about it and can afford to, dispose of your mattress.

Insure that you mark the mattress with its condition and do not allow mattress deliveries to pick up the old one as this can contaminate others.

Place them inside of plastic and mark them before transporting them to keep from spreading the bed bugs around.

Finally, you will want to wait on having new mattresses delivered to your home until after the chemical treatment is completed.

<u>Step Five: Treatment At Last</u>

When the exterminator does come to your home, he will take the time necessary to go through the entire thing and treat it. At least, he should do this. He will use two types of products.

One is called an ***instant kill***. This pesticide will kill the bed bugs as he finds them. He may also use this product in large hiding spots that he finds.

The second product that the exterminator will use is that of a barrier. When the product is placed around the home, in and near hiding places, the bed bugs will cross over the barrier and die because of it.

This product can last up to 60 days although it is most effective the first two weeks.

It is important for the professional to do a thorough job in the home. This is what will determine how effective the treatment is.

After the treatment is done, you may still notice one or two bed bugs around. This is common and as long as their numbers are low, you can assume that the treatment is working. It can take up to two weeks to fully kill off the bed bugs.

During this time, monitor them. If you notice additional bed bugs, you can have an additional treatment done.

You should not vacuum the area in which the treatment, especially in the bedroom, has been done. This can remove the barrier too soon. Your professional will give you a proper timeline here.

Lastly, it is important to note that sometimes we may feel that we see, fell and notice bed bugs when they are not around. As we mentioned before, bed bugs can cause us to be delusional. Some even feel as if the bugs are following them from place to place.

With proper treatment of the home, you can safely remove bed bugs from the home altogether.

<u>Chapter 10</u> - Preventing Them

You have spent time and money getting rid of the bed bugs in your home. Now, you need to keep them out of it. While there is never a guarantee that they will never come back, there are several things that you can do now that will help you to keep them from returning.

Here's a checklist for you to use to keep bed bugs at bay:

- ✓ As hard as it seems, try to keep bed bugs from returning by keeping yourself out of infested homes, hotels, motels, dorm rooms and other locations we mentioned that they like to breed.

- ✓ Inspect all of your clothing, luggage and other products that you take with you on a trip. Make sure to look for the signs of bed bugs that we mentioned including what they look like, the blood stains and the fecal stains.

- ✓ If you bring anything into your home that is second hand, make sure that you wash it and take great care in inspecting it. This including things like furniture, bedding, mattresses and box springs. Always do a detailed inspection before it enters your home.

✓ If you live in an apartment complex, talk to the landlord about their efforts to keep them at bay. If you have treated your apartment but the entire building was not treated, it is likely that there will still be potential problems in the future. If you are moving into a location, ask if there has been any occurrence of bed bugs and what is done about it.

✓ Keep crevices, holes, cracks and other pathways into your home caulked. Check them yearly to insure that they are fully sealed.

✓ If you have birds, pets or rodents in the home, it is important to realize that bed bugs will also use these hosts if they can not find a human host. Insure that any pets are treated (they are often mistakenly treated for fleas.)

<u>Chapter 11</u> - Sanitation and Bed Bugs

We talked about the fact that bed bugs are not in homes because people are dirty. Poor hygiene of a home is not likely to be a reason for bed bugs to arrive. They will not, in fact, come into your home because it is messy. But, sanitation can be effective at preventing them as well.

Your home does not need to be messy to have bed bugs, but if you use proper sanitation, it is harder for them to penetrate into the home.

If you have had a chemical treatment in your home recently to treat bed bugs, make sure that you follow the specific directions provided here as well as from your professional exterminator as to what you can and can not do.

Keeping them gone is a very important job for you to do. Here is another checklist of things that you can do to help you to keep the bed bugs any place but in your bed.

✓ Keep your bedding and clothing clean. When you launder them, make sure that they are washed in the hottest water permissible. Allow them to remain in the dryer for at least 20 minutes on a medium to high heat. This will kill any lurking bed bugs.

✓ Vacuum your mattress monthly. This means removing coverings, vacuuming the area around the mattress, the bed frame, the crevices and the carpeting underneath your bed frame effectively. Dispose of the container of product that you remove into the trash outside of your home.

✓ Do not try to steam clean your mattress. It is very hard to dry a mattress completely. If you do not accomplish it, there will be problems with things like mold and mildew. Then, the mattress will be a loss completely.

✓ If you properly clean the mattress, you will need to wrap it tightly. You can do this in several ways. First, you can use plastic to encase the mattress completely. This works well on the box springs.

✓ Second, you can wrap it in a high quality dust mite controlling cotton covering. These are designed to keep anything from getting in or getting out.

✓ If there are bed bugs in your mattress, they will perish as long as you do not remove this covering for at least one full year. If there are bed bugs other places, they will not be able to get into the mattress.

✓ Use a vacuum to keep the carpeting clean. You should vacuum the home as often as needed, but at least weekly to help you to pull out any bed bugs that may be lurking.

✓ Make sure to use the attachment to get into all crevices and cracks that may be an ideal location for the bed bug to present itself.

✓ Bed bugs may be lurking in your car, in your furniture or other places that you spend a great deal of time. In order to properly remove them, you need to use chemical treatments as necessary.

✓ Keep them clean, then, by vacuuming and monitoring them for signs of re-infestation of the bed bugs.

✓ If you feel that the treatments for bed bugs have not fully worked, talk to your exterminator about what you should do. Again, cleanliness can help, but in your cleaning, you may also find additional infestations. Monitor them carefully. Always have a look out for bed bugs that may be coming for a visit.

✓ Finally, make sure that you monitor your children as well. Children often visit friends, family or other people. They can even come in contact with a friend at school that has bed bugs in their clothing.

If you notice the tell tale signs of skin irritations or bite marks, it may be that bed bugs are lurking in her bed. Make sure that you remove and treat anything including stuffed animals for bed bugs.

<u>Chapter 12</u> - What to Know And Do

It bears warning that bed bugs are going to make a come back and are likely to infest more homes in North America quickly. While this happens, there are likely to be many products on the market that may claim to help. There are likely to be many that do not as well. Take the time necessary to really check them out, first.

It pays to be vigilant when it comes to bed bugs. Always look for signs of them. While it may not seem like anything to worry about, one of the most harmful effects that bed bugs can have on humans is the emotional and psychological stresses that it causes.

If you are feeling as if you can not get the bed bugs off of you, or that they are everywhere, realize first that this is a common problem that people face when they are dealing with bed bugs.

If you are in this situation, it can be wise to talk to your doctor about this anxiety, fear or problem as it can escalate into something much more. There are treatments that can help you to overcome these types of problems with bed bugs.

Self Treatment

If you go back to our treatment section in this ebook, you will notice that we took the professional cleaning way to go. We did this because it is by far the most successful of treatments, especially when you work with a skilled professional exterminator that has worked with bed bugs before.

But, if you feel more comfortable with handling bed bugs on your own, by all means you can give it a try. The most important things to do are included in the pretreatments listed in that chapter.

You should also use pesticides and other products that are both safe to you and to your pet when treating bed bugs. There are insecticides on the market that can help you to do you own self treatment of bed bugs. How well they work is really undetermined as of yet. Some are sprays others are powders. Choose products that you know will provide you with the highest level of effectiveness.

Conclusion

It is a long and hard process to get rid of bed bugs. These animals, although almost harmless to the human body, can be detrimental to our own well being emotionally. Therefore, we strive to provide ourselves with a safe and clean home.

Unfortunately, even those that do practice the safest matters of prevention from bed bugs will find that they are infested. While it is hard to believe and devastating to hear, the infestation of bed bugs does happen quite often.

It is even worse to realize that bed bugs are likely to be infesting more homes across North America. To take this to the next level, the inexperience of exterminators to handle these types of infestations can be downright nerve-racking.

Yet, with a bit of skill, lots of hard work, and just the right timing, you can get rid of bed bugs once and for all.

www.ingramcontent.com/pod-product-compliance
Lightning Source LLC
Chambersburg PA
CBHW062057280526
45788CB00003B/1270